I am fascinated by space,

how it can affect our
activities and perception.

Mistakes are part of the process.

As soon as one enters Augustas Serapinas' solo exhibition Wood and Snow at Galerie Tschudi in Zuoz, the first association that arises is that of the meaning of its title. The show takes place in the winter of 2022-2023, the coldest and darkest time of the year, when the streets and buildings keep getting covered with a new layer of snow; the gallery, like every home, becomes a refuge from the harsh winter, where as soon as you step into the ground-floor hall you are greeted by Serapinas's wood works installed in the rustic interior of the space. These works are fragments of walls of abandoned and decaying wooden houses, with windows, shutters, doors and other usual attributes. They blend almost naturally into their surroundings in a submissive harmony with the gallery's wooden ceiling and other wooden details. One of these found structures divides the space lengthwise into two parts, while the others cling to the original stonework walls by covering them. The arrangement of the objects narrows the perspective of the gallery corridor and leads the eye towards a wooden object at the very end, which upon approaching turns out to be a real wooden tiled pitched roof of a shed.

One floor up, separate structural elements of wooden buildings—decayed beams, rusted window glasses, decorated shutters and a rectangular fragment of a burnt roof, more than 6 metres long and almost 2 metres high—are displayed. In the rear room of the first floor, a former hay barn space, the largest work in the exhibition is installed— a structure of wooden walls and their connecting beams, almost 4 metres high and covering a perimeter of 5 metres in length and width. Although it is quite obvious that this structure is an improvised, monumental sculpture and has neither a practical function nor inherent qualities of inhabited architecture, its materiality inevitably prompts us to imagine it as a building preserved in a state of limbo between not-yet-built and half-demolished. The ambivalence of the experience of this object is reinforced by its organization in space: on the one hand, touching the curved beam of the building, it becomes part of the building's structure and ties the whole gallery space together like a knot; on the other hand, this touch creates the impression that the weight of the beam is pressing down on the structure, thus adding fragility and drama to its monumentality.

As is characteristic of his work, Serapinas refers to the works of Wood and Snow by their real names—e.g. House Wall from Didžioji Kuosinė, Roof of a House from Steponių Village, or Window Glasses from a House from Skirgiškės. The artist looks for these buildings—some of them dating back to the 19th century—in different Lithuanian villages by browsing popular classifieds portals, where they are for sale as fire wood, on condition that the buyer dismantles the building and takes it away themselves. The exhibition is a continuation of Serapinas' long-standing interest in the relationship between vernacular architecture and modern monumental sculpture. The first project exploring this theme was Four Sheds (2016, Fogo Island Arts, Newfoundland, CA), where the artist, together with other members of the local community, formed the exhibition display from huts of local fishermen (where they store their wood and fishing gear and trim the fish they catch), lined up along the shores of Barr'd Islands and doomed to be demolished. Serapinas transformed these buildings according to the characteristics of the Fogo Island gallery space and, after the show, turned them into a transport container for future trips to other exhibitions. One of the most recent projects on the theme of vernacular architecture was implemented at the 13th Kaunas Biennial (2021), where Serapinas placed a real ancient Lithuanian wooden sauna in a prestigious, representative part of the city, at the confluence of the Nemunas and Neris rivers. The building, strangely distinguished by the lack of windows or other openings, transformed vanishing architecture into a ghostly monumental sculpture in the living fabric of the city.

By buying up centuries-old buildings no longer fit for residential purposes, Serapinas not only saves them from turning into ashes and smoke, but also opens them up to endless possibilities of formal and structural variations. In each exhibition, the buildings are reshaped to correspond to a new display context and always create a meaningful tension with the materials in the environment as well as the ways they are processed and used. The artist meticulously cleans the decayed wood of the walls and beams, revealing both the usually invisible handcrafted elements of the building—the structural nodes and the organization of details— but also the true extent of nature's destructive effect on the materials. The decay process is stopped by proofing the cleaned parts of the building with natural antiseptic materials or traditional techniques, such as treating roof segments with fire—a technique previously used to disinfect the wood and protect it from pests.

Serapinas manipulates not only the parts of the building by creating new forms, but also the interplay between materials: the main structure in the former hay barn space is built without any nails and these are used to smoke the window panes by covering them in rusty dust of the finely grinded original nails and then re-firing them. By exposing traditional handiwork in a decaying building and highlighting nature's destructive effect on the material, Serapinas surrenders to this tension. Rather than shaping the material by creating a resistance to it, he searches for new forms and meanings by working together with existing resistances. This sparseness of the decaying materials and interventions in them has a clear ecological subtext and comments on the principles of the artist's work: the modesty of the works and the simplicity of creative gestures breaks down the usual distance between the artwork and the viewer, highlights the tactile qualities of the works and appeals to touch through a practical everyday approach to the human relationship of the objects.

The semantic tension encoded in the title of Wood and Snow unfolds as a more general principle that threads through much of Serapinas' work.

This dynamic between revealing and concealing manifests itself as a laconic and critical rethinking of a particular place through the specificity of materials and their interplay, emphasizing what is essential to the identity of a place yet usually remains invisible or purposefully hidden from the outsider's eye. An emblematic example of this principle is Serapinas' early and little-known work in the public sphere—an intervention into the Warsaw market, developed in the Dom Kereta residence in 2013.

In the few publicly available photographs of the work, we can see the twenty-something artist wearing a hi-vis acid green vest and, together with an older colleague, patrolling the aisles of market stalls. In Warsaw, as in other post-Soviet cities of Central and Eastern Europe, the market is still perceived as an exception to the representative order of the city, or as a multifaceted medium and symbol of illegal goods and the shadow economy. As Serapinas says, 'several social spheres are clearly connected in this place, cleverly covering each other, and this system of concealing and revealing is sufficiently developed and creative to be discussed in an artistic context.' Serapinas used the security guards' changing booth space to hang a holographic image he found in the same market. The image, when moved, reveals either a half-naked girl or the Eiffel Tower and a luxurious vintage car in the background of a Renaissance palace. According to the artist, the hanging of the picture was the real initiation into the ranks of the market's security guards, and, apart from a few photographs, it was also the only gesture of a long-term, tangible nature documenting this work.

Serapinas' intervention in the Warsaw market clearly reveals that what he calls 'the system of concealing and revealing' begins with human relationships and, whatever final form the work or exhibition takes, always marks them. Serapinas' work is characterised by seeing institutions through the invisible people who work in them and the importance of their work. At the same time, he rethinks, within a given context, how different temperaments, knowledge, skills and, most importantly, life experiences create a unique inner climate that goes beyond human relationships and invisibly begins to affect objects, their circulation, specific works and even entire exhibitions. This is why different works are often dedicated to particular people. This can be seen in projects such as Georges at the Museum of Contemporary Art M HKA (Antwerp, 2014), where the artist, on the suggestion of the museum's technical manager, demolished a wall in the exhibition space and extended it into a shaft in the building, unknown to the visitors and even to the staff. Another example is Sigi (exhibited on the terrace of the Kunsthalle Wien in 2017), a sculpture of a cat made by children out of a cardboard box which the institution's chief financial officer found discarded on the street and placed in her office. The principle of foregrounding invisible, culturally insignificant work or activities that are not even considered real work or creative

production also guided the 2019 exhibition February 13th at Emalin Gallery in London, where the artist brought snowmen made by anonymous Lithuanian children and exhibited them in the gallery room turned into a freezer. In the context of these previous projects, vernacular architecture reveals itself both as a marginal, vanishing material heritage and as a narrative about communities, their different generations, their relationship to their environment and the skills that remain alive in the constant renewal and creation of their most intimate material environment.

Serapinas reminds us that materials do tell stories. Although the artist's use of parts of huts as ready-made building materials to make abstract sculptural compositions bears obvious similarities to the practices and formal experiments of the sculptors who created in the 60s and 70s, such as Mark di Suvero, Antony Caro, Donald Judd, Alice Aycock, or even his teacher, Lithuanian Mindaugas Navakas, the need to use these abstract compositions to encode cues for the development of a narrative seems to be important as well. For example, the artist tells us that the well-preserved roof of the building indicates that it used to be covered with asbestos shingles (also known by the slang term 'shifer'), a material popular in the Soviet era, which is poisonous when crushed and explosive when heated. This simple upgrade is a silent indication that the house was inhabited by several generations of people.

The narrative elements in the exhibition weave a fragmented story about the current state of life in the periphery of Lithuania and the region. During the three decades of Lithuania's independence, a quarter of the country's population emigrated, abandoning not only individual homesteads, but entire settlements and villages, which were thus left to decay. However, the landscape of Lithuania's provinces has been transformed even more by the return of these emigrants, with better financial conditions encouraging them to actively renew their material environment, thus simply eliminating the outbuildings and residential architecture that had been created, maintained and preserved generation after generation. In the exhibition, Serapinas has captured this change in Lithuania's (and perhaps the whole region's) landscape and material culture as a change in the techniques of communal memory. The title, Wood and Snow, is a simple and direct reference to Zuoz's landscape and the materials in the exhibition, which, in harmony with the gallery space, offer the viewer endless associations arising from the interplay of tactile and visual signals. At the same time, it can also be read as a poetic metaphor for the ephemeral nature of human life and memory. The wood used to build oneself a home will inevitably decay with the passage of time, and its remnants will merge with the horizon in the whiteness of the all-embracing snow. What comes to mind consequently is urbanist Paul Stangl's idea that in the study of memory culture, vernacular objects and architecture are usually linked to community memory, whereas monuments are associated with the representation of abstract community values. The forms of vernacular architecture are firstly determined by basic human needs, with representation playing a secondary role, whereas monuments can be considered as almost excl-usively representational objects—professionally programmed containers of memory. By appro-priating objects of vernacular architecture and remodelling them in ways that are close to formalist monumental sculpture, Serapinas combines these aspects of the representation of living memory and abstracted values into a complex whole, in which anonymous vernacular material practices are given the attention usually given to 'high' art, while the often socially hermetic and elitist artistic production is given the traits of everyday practical human activity.

The melancholic simplicity of Wood and Snow also reveals a quiet irony characteristic of Serapinas' work. After all, what are fragments of neglected buildings from the geographic margins of Lithuania doing in a small-town gallery in the heart of the Swiss Alps? However, by making use of the infrastructure of contemporary art production and dissemination as well as its usual operating principles, Serapinas creates not only a living model of the preservation of the material culture of the past, but an entire dynamic practice of making it relevant and seen again, which is hard to imagine being implemented outside the contemporary art sphere.

**Every site, every situation
is a unique set of things.**

without showing any architecture.

The house turns into sculptures,

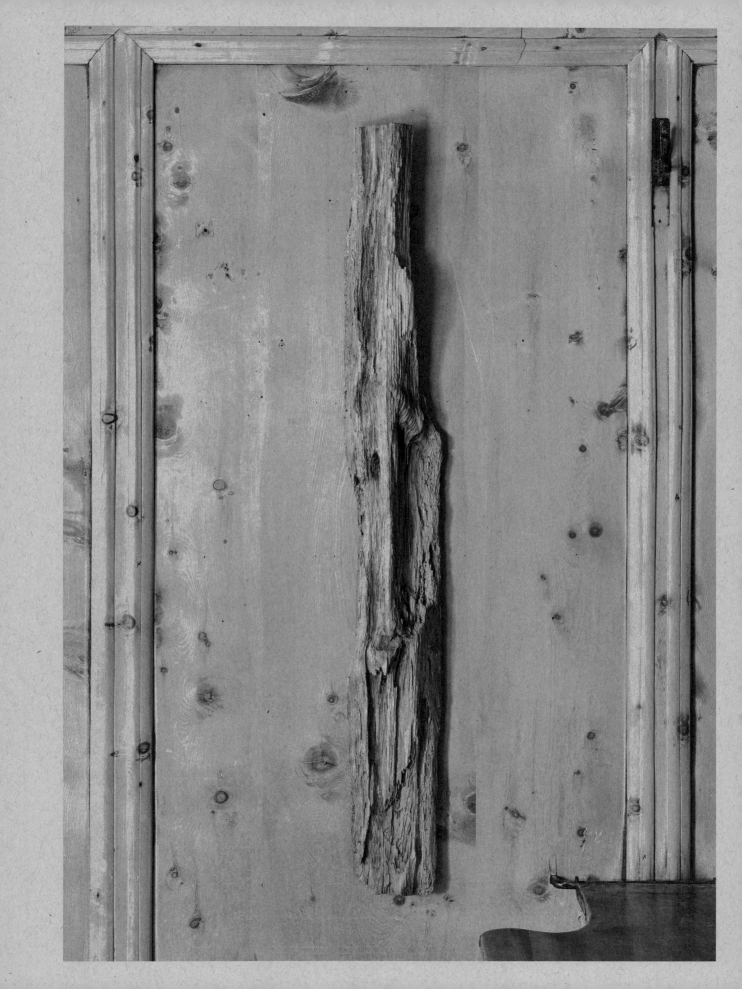

and each sculpture stands for the regional wooden architecture.

**Secret and public is
a matter of perspective.**

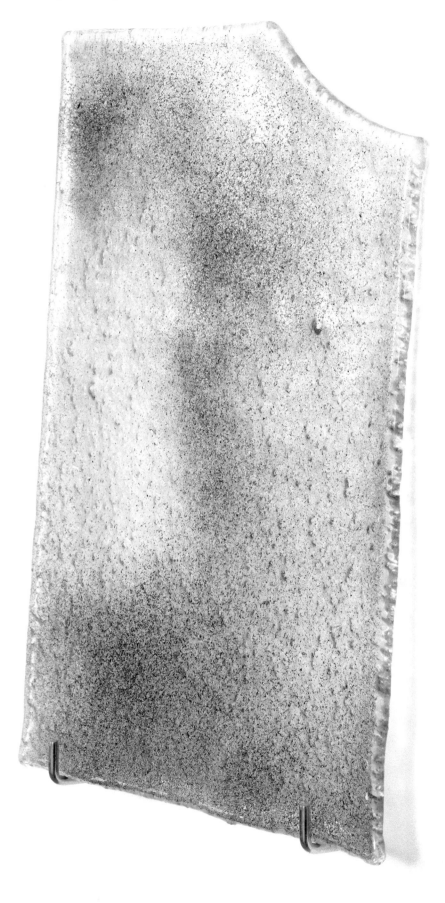

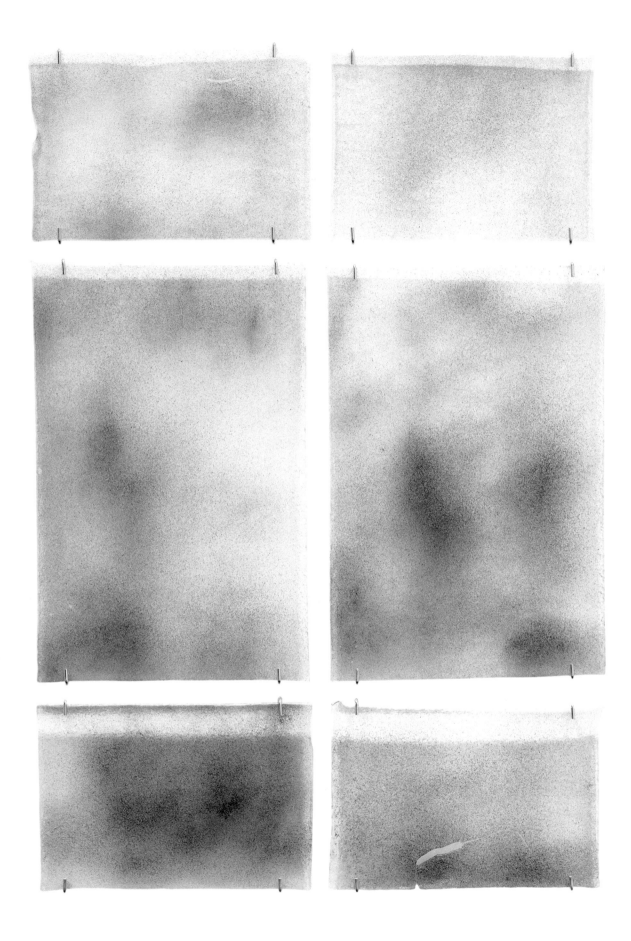

Aesthetics as continuation of thought.

A shift between the perspective of the artist and the viewer.

Selected exhibitions
2022 – 2013

In 2022, for his first exhibition Five Stoves at Galerie Tschudi in Zuoz, Augustas Serapinas acquired five former tiled stoves dating to the late 18th, 19th and early 20th centuries. He converted those essential parts of domestic interiors into contemporary objects of utility and thus directs the focus to the social and political changes in domestic living space.

For his exhibition Roof and Walls at Emalin in London in 2022, Augustas Serapinas obtained a dilapidated wooden house located in the southeastern village of Rūdninkai and turned his attention towards the shingled pine roof of a residential house.

For steirischerherbst'22, Augustas Serapinas commented upon the infrastructures of Neue Galerie Graz and the importance of hideouts and the fraught lines of communication in times of war. Small openings in museum walls serve as vents and windows for small hiding places. For the duration of the performance, these spaces are inhabited by museum employees Margit, Maya, and Vanessa, after whom the piece is named.

At Art Basel Parcours in 2021, Augustas Serapinas placed his Mudmen in the UBS branch in Basel, first exhibited at Riga International Biennal of Contemporary Art 2 in 2020. Thinking about new ways of creating snowmen, this humorous yet critical installation raises questions about global warming.

In <u>Summer</u> <u>in</u> <u>Rūdninkai</u>, on view in <u>2021</u> at <u>Apalazzo</u> <u>Gallery</u> in Brescia, Augustas Serapinas presents works of an architectural nature based on the so called vienkiemis: traditional wooden buildings erected in the 1920s–'30s in the rural areas of Lithuania.

The works shown at the exhibition at <u>the</u> <u>Center</u> of <u>Contemporary</u> <u>Art</u> in Tel Aviv-Yafo in <u>2021</u> are a result of Augustas Serapinas' exchange with the people in the surroundings. <u>Diana</u>, one of the staff members of CCA, was not only the artist's inspiration for the exhibition title, but also for three objects, produced by local artists, artisans, and fabricators.

For <u>Kunsttage</u> <u>Basel</u> <u>2020</u>, Augustas Serapinas collaborated with Swedish craftsmen to build a 20-meter-long and four-meter-high wooden sculpture <u>Standtune</u> <u>for</u> <u>the</u> <u>Square</u>. Placed in public space, the work, is inspired by traditional defensive walls and military demarcation objects.

Under the circumstances of a lack of snow, together with residents from Riga Augustas Serapinas built <u>Mudmen</u> of soil, water, and hay for the <u>RIBOCA2</u> in <u>2020</u>. His bulbous, grotesque, yet endearing figures are a reminder of climate change and draw our attention to the kinds of rituals we might have to follow to replace our traditional customs.

In <u>2020</u>, Augustas Serapinas mirrored the adjacent building site into the exhibition space <u>P/////AKT</u> in Amsterdam. With this facsimile of the yet-to-be-constructed five-story building, named <u>20 Apartments</u>, the artist draws the viewers' attention to the consequences of gentrification.

For the <u>Various Others</u> show in Munich <u>2019</u>, Augustas Serapinas displayed a fragile skeletal structure of a weathered <u>Greenhouse from Užupis</u>—an area in Vilnius going through vast gentrification.

In his pursuit of creating alternative points of view, at the 58th Venice Biennale in 2019, Augustas Serapinas placed his Chairs for the Invigilator throughout the Arsenale space. The lifeguard chairs were meant to be used only by the exhibition invigilators who were not supposed to sit down while supervising.

In February 13th, Augustas Serapinas exhibited a group of snowmen in the gallery space of Emalin in 2019. The journey of the snowmen—created by children in Vilnius Vingis Park on Lithuanian playgrounds—began on the date of the exhibition's title, when there were loaded into a van and driven to the United Kingdom, where they rested in a freezer at the gallery during the exhibition.

The work <u>Vygintas</u>, <u>Kirilas</u> <u>and</u> <u>Semionovas</u> by Augustas Serapinas, which was on view at the <u>Baltic</u> <u>Triennal</u> <u>13</u> in <u>2018</u>, consists of fragments from the wall of a decommissioned nuclear power station. The work is named after the children of former workers of a disused nuclear power plant in Lithuania, with whom the artist worked to build abstract sculptures.

In the process of preparing his exhibition at <u>Basement</u> <u>Roma</u> in <u>2018</u>, Augustas Serapinas became an observer of the daily life of the people and their dogs at Pretty Pets, a small pet salon next to the exhibition venue. The exhibition <u>Where</u> <u>is</u> <u>Luna?</u> became a metaphor for a society that the artist investigates with an anthropological approach.

Waiting for Another Time is a reflection on the concept of preservation. During the preparation of the exhibition for Apalazzo Gallery in 2018, Augustas Serapinas visited Chiesa di San Silvestro in Folzano and was astonished by the fact that the fresco was removed from the church walls. Researching the history of the church, he found parts of the original frescos at a local store studio and displayed them at the gallery.

Developing a relationship through production and an exchange with two neighboring businesses, Augustas Serapinas collated a collection of memories from the family-run business Clow Group and David Dale Gallery in Glasgow, where the exhibition Blue Pen took place in 2018. The works shown are interpretations of these memories.

At <u>1857</u> in Oslo, Augustas Serapinas showed <u>2017</u> <u>Gym</u> (Jõusaal), a restaged work initially conceived in residence at the Estonian Academy of Arts in Tallinn, Estonia. The gym equipment, he recreated from memory and made from materials found on the site of the exhibition venue.

A casket that mimics the shape of a cat in a childlike way, which Augustas Serapinas found in the staff office of the <u>Kunsthalle</u> <u>Wien</u>, became his model for the large-scale installation named <u>Sigi</u> in <u>2017</u>.

In <u>2016</u> at the <u>Fogo Island Gallery</u> in Canada, Augustas Serapinas encountered a local shed that was soon to be demolished. While dismantling it himself, he learned from locals that it was altered three times before and therefore titled his work <u>Four Sheds</u>. His interest in abandoned vernacular architecture evokes a history of materials, techniques, and traditions, and the human processes embedded within them.

Augustas Serapinas' first exhibition <u>Housewarming</u> at <u>Emalin</u> in <u>2016</u> highlights the problem of gentrification in East London. His installation is a smaller replica of the building, the gallery temporarily occupies and tells the story of the former locksmith evicted from this very site.

As the third resident in <u>DRAF</u> Studio in <u>2016</u>, Augustas Serapinas researched the history of DRAF's Camden building in London, built in the 1870s as a furniture factory, carefully examining the current space for traces of former identities. For his installation <u>Dusting the Ground</u>, he used scraps left over from past DRAF projects to create new woodwork compositions.

In <u>2015</u>, the exhibition <u>Philip</u>, <u>Lukas & Isidora</u> uncovers the social interactions that might take place in the apartment building that towers above <u>SALTS</u> in Birfsfelden. For the installation, Augustas Serapinas borrowed various elements from the surrounding apartments and re-arranged these in the storage back rooms of the exhibition space.

Augustas Serapinas' contribution to the 6th Moscow Biennale in 2015, Behind the Third World, is situated behind Qiu Zhijie's painting called 'The Map of the Third World'.
It is a hidden tearoom, constructed of leftover materials from the exhibition venue.

In 2014, Augustas Serapinas has collaborated closely with Georges, the chief technician of M HKA in Antwerp to identify hidden, forgotten and unused spaces in the museum. The resulting work is a cavity behind the exhibition halls on the ground floor, which he has made accessible to the public.

The starting point for the 2013 residency program at the <u>Dom</u> <u>Kereta</u> / <u>Keret</u> <u>House</u> in Warsaw was the theme <u>Spaces</u> <u>of</u> <u>Emptiness</u>. Augustas Serapinas took an anthropological approach to the topic and engaged with the guards of the nearby market halls Hala Mirowska.

The exhibition <u>Jakubowska 16/3</u>—<u>Jakubowska 16/3</u> at the gallery <u>BWA</u> in Warsaw in <u>2013</u> refers to the location of the gallery. Augustas Serapinas has taken on the history of the building and the history of its inhabitants and transformed it into a site-specific installation.